REFLECTIONS
of a
PRODUCT
ENGINEER

Order this book online at www.trafford.com
or email orders@trafford.com

Most Trafford titles are also available at major online book retailers.

Print information available on the last page.

ISBN: 978-1-4120-8409-3 (sc)

Our mission is to efficiently provide the world's finest, most comprehensive book publishing service, enabling every author to experience success. To find out how to publish your book, your way, and have it available worldwide, visit us online at www.trafford.com

Trafford rev. 11/20/2018

 www.trafford.com

North America & international
toll-free: 1 888 232 4444 (USA & Canada)
fax: 812 355 4082

CONTENTS
CHAPTER 1 LAYING THE FOUNDATION

At an early age McIntosh displayed an interest in design, an example of which is an experiment harnessing the earth's magnetic field. After high school and a degree course in mechanical engineering he was employed briefly in product development. His keen interest in product design prompted him to accept a scholarship to The Institute of Design in Chicago.

On staff were some of the original instructors from the German Bauhaus, from which the institute had evolved. While studying for his M.S. degree McIntosh performed a fully described task for guest lecturer Buckminster Fuller.

CHAPTER 2 PHILOSOPHICAL DESIGN PRINCIPLES

The practice of industrial design fifty years ago was more difficult, first because of its novelty, and also because all drafting was done manually. There were, of course, no computers with design capabilites.

An unorthodox approach to design, concentrating on one well developed concept rather than a myriad of sketches, is described; and the resulting necessity of presenting accurate perspective drawings is demonstrated.

CHAPTER 3 THE POLAR DESIGN GRAPH

This concept was developed to illustrate what constitutes good and bad design, and how they might relate to each other with respect to cost. Both are available at prices ranging from high to low. Design value and price are unrelated and mutually exclusive.

CHAPTER 4 ANCEDOTAL EVIDENCE

First there is some advice for students, then the relating of ancedotal experiences in the fields of industrial design and expert witnessing in court litigation. There is also reference to experience as an instructor in colleges and universities and the nature of the knowledge imparted to students.

CHAPTER 5 RETIREMENT

This chapter recounts some activites following retirement; but, mostly, it illustrates a retrospective exhibition of an illustrious design career.

CONCLUSION

There are no regrets in retirement for various listed reasons.

CHAPTER 1 LAYING THE FOUNDATION

It all started in 1939 when I was fifteen years of age and in high school. My elder brother was at Queens University in Kingston, Ontario, studying mechanical engineering. Being curious, I began examining the textbooks he brought home and became intrigued by electromagnetism, especially direct current machinery, electric motors. I began fabricating simple motors from the technology described in the books, using readily available materials. Speaking of available materials, my source of electricity was the old style of telephone batteries which were about six inches high and three inches in diameter. They were used in the old wall-mounted phones that had a crank, used to call the operator. Every year a man in a telephone company truck would collect the used batteries and install new ones. Instead of giving him our used batteries I would ask for and receive as many as forty used batteries from his truck (I don't know how I eventually disposed of them). He was glad to be rid of them. They still supplied enough current, whether connected in series or in parallel, to drive the motors. The armatures, or cores, of my motors were made of steel, though soft iron would have been better. They had copper strips for commutators and bar magnets to induce rotation. I don't remember what I used for brushes. In any case, after making a few motors in various configurations, I wanted to try something different and unique.

Realizing that the Earth has its own magnetic field and causes a compass needle to orient itself accordingly, I wondered if it would be possible for a motor to be driven by the Earth's magnetic field, rather than field magnets. I knew that it would require virtually no friction because of the minimal magnetic field force available. Such a motor would not provide enough torque to drive anything, but would demonstrate a principle.

After considerable analysis and design I successfully fabricated a motor that produced the desired result (Plate 1). It rotated at about 100 rpm. I used some steel strips, from a Meccano set, wrapped with insulated copper wire for an armature, and suspended it on a needle point. The depending wire ends (brushes) made contact with two arced pools of mercury, which served as a commutator. Why mercury? It has a positive meniscus rather than a negative one like that of water. The meniscus is the edge condition of a liquid where it makes contact with its container. Mercury forms a convex edge where it meets the container. Thus, in this case, the pending wires could pass from one pool of mercury to the other without striking the container, which was formed from plaster. We did not know in those days that mercury was poison and we played a lot with "quicksilver" back then. As a matter of fact when I placed bar magnets (broken horseshoe magnets from an old car magneto) near the armature it spun so fast that the mercury hit the walls of the room in small droplets. We will discover later that mercury was not the only now forbidden substance that I used in ignorance. It's a wonder that I have lived to attain octogenarian status.

A year later I took a summer job with a printing company in Paris (Ontario, that is). I had not yet competed high school and, while there, I considered entering the commercial art field; but the artist on staff quickly talked me out of it, suggesting I finish high school and continue my education. After meeting with the principal of Brantford Collegiate, accompanied by my parents, I decided to complete high school matriculation (grade 13) in Brantford. The following year,1942, I enrolled in mechanical engineering at the University of Toronto. You could say that the post secondary exposure was uneventful.

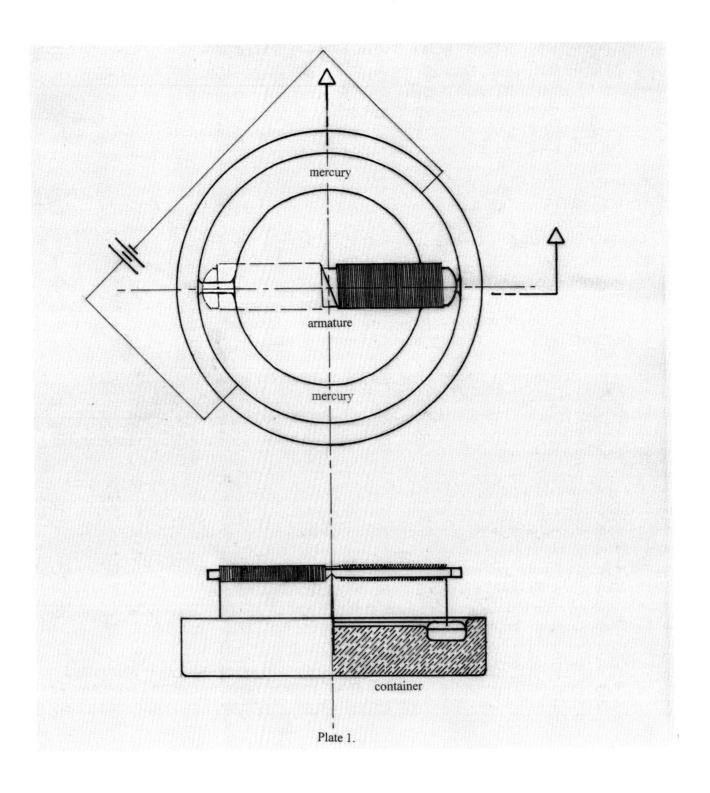

Plate 1.

Plate 1

On graduation in 1946 I accepted an offer from T. S. Simms and Co. Ltd. in Saint John, New Brunswick, to act as their Product Development Engineer. I began designing brushes, brooms and mops for the oldest brush company in the Commonwealth, incorporated in 1866. After two years I was informed by Professor McIntosh at U of T, one of my former professors (but unrelated), that scholarships were being offered by the NIDC, the National Industrial Design Committee, to practicing engineers or architects to attend the Institute of Design in Chicago and study industrial design. With my experience at Simms I knew, by this time, that industrial design had to be my future, and so I eagerly applied for and received a scholarship. Five of us were awarded scholarships and went to Chicago in 1948, four architects and myself, an engineer. Two years later Joan Robinson, Bill Greer, Murray Simpson and I received Master of Science degrees in Product Design from the Illinois Institute of Technology, since the Institute had been amalgamated with IIT in 1949.

The Institute of Design had evolved in 1944 from the New Bauhaus, founded in 1937 in Chicago. The original Bauhaus was established in Germany, but it was deemed subversive and dissolved in 1933 by the Nazi government. The former Bauhaus master, Moholy-Nagy emigrated to Chicago and continued as the head of the New Bauhaus and then the Institute of Design in Chicago until his death in 1946. Serge Chermayeff, international architect and painter, then became director, and was serving during our two year course. Among our instructors were Nathan Lerner, Eugene Dana, John Wally, Hin Bredendieck, Otto Kolb and Harry Callahan, the famous photographer. At least some of them had emigrated with Moholy-Nagy from the German Bauhaus.

For these recent immigrants language, particularly pronunciation and spelling, proved to be a minor problem. Otto Kolb was a rather young instructor, German, but probably not with the original Bauhaus. On his first morning with us he proceeded to assign a design problem on the blackboard by writing the date for completion as "Okt. 15ᵗʰ". Someone said that no, it is spelled with a "C", and so he changed it. Then he wrote the assignment as "scetches" and became confused when we said it was spelled with a "K". The humour of the situation broke the ice. Hin Bredendiek, on the other hand, gave us an assignment, verbally, to design a dishrag. At least that is what we thought. With the Bauhaus abhorrence of ornamentation and strict adherence to function, I could imagine a grey cloth, ten inches square, perhaps with rounded corners. But no, it was a dish rack that he wanted, a reasonable assignment, since automatic dishwashers had not yet come into their prime in 1949. Bredendiek later departed for Georgia Tech (a ramblin' wreck, I don't think so).

Our two year learning experience began with a condensed one semester foundation course which I found a difficult adjustment coming from an engineering background. We were experimenting with a wide variety of materials such as paper, wood, plastics, textiles, metal and plaster. Though we produced some interesting results, the paper folding seemed like going back to kindergarten. I was happy to get home for the Christmas holiday. However, as we proceeded beyond the foundation course to more complex static and dynamic assemblies and simple product design assignments I really began to enjoy the experience. I remember one mobile design that I fabricated. It consisted of a concentrically suspended series of rings, with small spheres attached, that, when motivated, would cause the spheres to trace orbits around the centre, much like planets around the sun or electrons around a nucleus, except that the orbits were random and not circular. The spheres were phosphorescent and the rings were black, so that, in a darkened room all that was visible was the group of spheres with their erratic motions. When I showed the mobile to Serge Chermayeff he insisted on keeping it, and so I never got it back.

For one semester in 1948 Buckminster Fuller was invited to the Institute of Design as a guest lecturer, and I was fortunate enough to attend his seventeen lectures on "Energetic Geometry". Bucky was a short stocky man with white hair in a brush cut. He was animated, personable and approachable. Instead of standing in front of the class to lecture he would often sit on a desk and dangle his feet back and forth. In his energetic geometry discussions he was mostly concerned about polyhedrons, multifaceted solids; but the skeletal forms were more his interest, as demonstrated by his Fuller Domes, for which he became famous, e.g. the American Pavilion at Expo 67.

Bucky's lectures included much more than geometry. He talked about his sleep experiments. He said that one day he was watching freighters unload at the dock and he noticed that the accessible surface holds were emptied much more quickly than the less accessible internal holds in the bowels of the ship. They seemed to take forever. This prompted him to think about sleep patterns. He wondered if we get maximum benefit from early sleep and decreasing benefit as sleep is prolonged. He decided to experiment by sleeping for half an hour in every six hours. He did this for three years and found that he could survive well on two hours' sleep in twenty-four. The problem was that he could not co-ordinate his life with societal standards, and so he terminated the experiment.

He also told us about his dymaxion car (see photo), first tested in 1933, which was "teardrop" shaped and had two forward driving wheels and one at the rear. He said that, if he were driving through a gateway or opening, he could turn hard right or left immediately the shoulders of the car, as well as his own, passed the opening. This was because steering was at the rear wheel and the curved and tapered side of the car would sweep past the opening without a scratch. He mentioned too that on one occasion he was driving (probably speeding) along a highway when the troopers spotted and chased him. He made a U-turn at speed and lost them.

Bucky also thought a lot about horizontal movement, rapid movement. He said that if a person in a vehicle traveled fast enough in a horizontal direction he could avoid falling to earth and remain in a horizontal path. He was describing orbiting but didn't actually use that term – definitely years ahead of his time. We are talking about a lecture in 1949, eight years before sputnik. And speaking of speed of travel, he drew a graph on the board with a horizontal axis at the bottom representing time and a vertical axis at the extreme right representing the speed of light. Then he inserted a hyperbolic curve hugging the horizontal axis at the extreme left and rising gradually as it approached the vertical axis, then swooping up to approach and eventually hug the vertical axis. This curve represented the

ncreasing velocity of the human being from walking or running at the extreme left, to horseback riding, to automotive and train travel, to air travel, passing the speed of sound, to orbital travel, to escape velocity. At this point, he remarked, we are on the cusp of the curve and eventually will approach the speed of light. The problem with this graph is that, as time marches on, we soon reach the vertical axis. Then what?

I was greatly taken with Bucky and his lectures. I am sure he sensed this and he asked me to perform a task for him. The following is a report prepared in 1949 describing this task. The drawings are recent.

"The following report describes briefly an investigation which was launched as a result of the observation of certain omni-directional phenomena, peculiar to the geometric form known as the dymaxion (Plate 2, Fig. 1).

"DESCRIPTION OF MOTION"

"The particular motion investigated involves simultaneous rotational and oscillatory component motions oriented about four separate axes passing through a common center.

"On examination the dymaxion appears to be composed of eight equilateral triangles and six squares (Plate 2, Fig.1 & Plate 3, Fig.1).Triangles are inherently rigid forms, whereas squares display considerable freedom of movement if the four line elements are left free to pivot in any direction at the four joining points. Thus we find that if the dymaxion is constructed by using straight rigid members for surface lines which are allowed to pivot freely at the joints, then the result will be a flimsy structure of which only the triangles remain rigid. In its most expanded form the structure is of dymaxion shape. Then if the triangles are rotated (Plate 3, Figs. 2 and 3) and simultaneously urged toward a common centre the structure assumes an octahedron shape (Plate 2, Fig. 2 and Plate 3, Fig. 4), each surface line being formed by two of the rigid members.

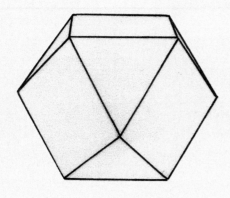

Fig. 1

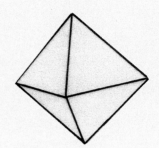

Fig. 2

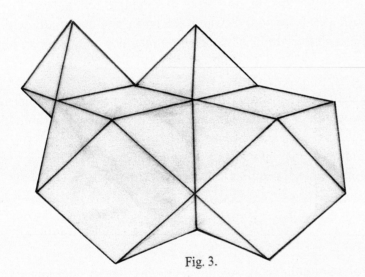

Fig. 3.

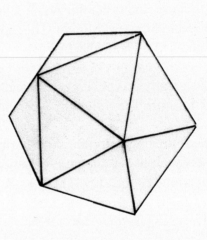

Fig. 4

Plate 2.

"

Plate 2

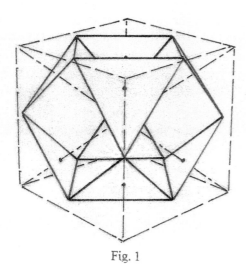

Fig. 1

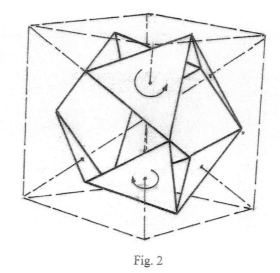

Fig. 2

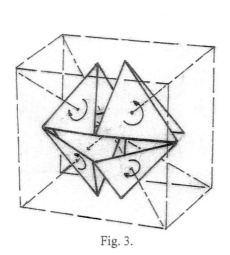

Fig. 3.

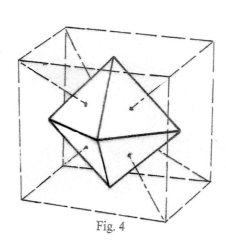

Fig. 4

Plate 3.

Plate 3

"Since the triangles are rigid we may fabricate them from a sheet material and achieve the same result as with the individual members. A further aid which keeps the structure symmetrical is to pierce holes in the centre of these triangles and mount them on wires stretched between opposite corners of a cubical structure (Plate 3, Fig. 1). If two triangles are mounted on each diagonal wire, one on each side of the cubical volume, and the vertices of adjacent triangles are joined with thread, then we have an excellent device to illustrate the dynamic principle involved. By grasping one of the triangles and turning it in either direction we find that the opposite triangle, on the same wire, turns in the same direction at the same speed because of the simultaneous motions of the intervening six triangles. If the triangles turn far enough, the edges of adjacent triangles become coincident and they form an octahedron (Plate 3, Fig. 4).By turning the triangles in their opposite directions we again reach the dymaxion stage and with further turning and contraction the triangles again assume an octahedron shape, with each triangle in a position 120 degrees from its former octahedronal position.

"The primary aim of the investigation was to construct the device so that the triangles, having assumed a new octahedronal position, could continue to be turned in the same direction indefinitely, causing the complete structure to pulsate alternately between the dymaxion and octahedron shapes. One complete cycle would consist of three complete pulsations, at which stage each triangle would have turned through 360 degrees and would occupy the same position as where it started. It is evident that mechanical connection of the vertices of adjacent triangles, as described above, would not permit continued turning in the same direction; and therefore some other way would have to be found to hold the triangles in their proper relationship to one another.

"The solution was found with permanent magnets. Twenty-four three inch magnets were obtained and fixed to the sides of eight plastic sheet triangles (Plate 4). The magnets were placed in rotation around the triangles with the magnetic flux "flowing" in the same direction in all triangles. Thus each triangle had a north and south pole at each vertex, which poles were attracted by dissimilar poles on each adjacent triangle. This experiment was therefore successful because all triangles tended to adhere to adjacent triangles at all times, no matter how many times they had been rotated from the original position."

I well remember demonstrating the device in front of the class. Bucky was ecstatic. Of course he expressed a desire to proceed in future toward some kind of motorized system that would operate with continuous pulsating motion. However this was the end of the experiment, as far as I was concerned.

I should mention that Plate 2, Fig.3 shows how space can be filled with dymaxions and octahedrons, with the latter occupying the interstices (intervening spaces) between the dymaxions. The two polyhedronal forms exchange places back and forth as the pulsating effect continues throughout space. In the two end states, where we have dymaxions and octahedrons occupying space, each dymaxion is surrounded by eight octahedrons and each octahedron by eight dymaxions.

It is also interesting to note (and I am not sure that Bucky realized this) at the halfway point toward collapse of the dymaxion to an octahedron (Plate 3, Fig. 2) or the expansion of an octahedron to a dymaxion, the individual structure passes through a stage where, with the addition of six struts, it would become a twenty sided figure called an icosahedron (Plate 2, Fig. 4). Of course the addition of these struts would be like crystallizing the whole space structure into a static rigid system consisting of triangles.

The question is, as with many of Fuller's concepts, does this whole study have any future or significance? I do not know the answer.

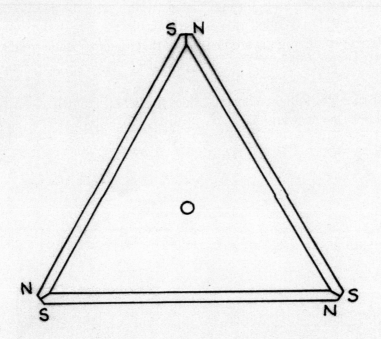

Plate 4.

Plate 4

CHAPTER 2 PHILOSPHICAL DESIGN PRINCIPLES

After my two year stint in Chicago I returned to Ontario and spent a year or two with one of Canada's first industrial design companies, John B. Parkin Co. Ltd. – not to be confused with the much larger sister company, John B. Parkin Associates, the architectural firm. It is interesting to note that one of the associates was John C. Parkin, well known for many of his accomplishments. John B. and John C. were not related.

Since I was doing all of the promotional and design work for the company myself, as the only employee, I soon launched out on my own and never worked for anyone else again. I was twenty-eight. Though I didn't plan it that way, I ran a one man business for forty-two years, occasionally engaging one, two
or up to five assistants on a contract basis; but 90% of the time I was alone. I doubt that that would happen, or even be feasible today. There are so many new factors and disciplines involved now that a team approach is a virtual requisite. I did, however, restrict myself to product design itself and never dabbled in packaging, graphic design, corporate identity, or other peripheral activities. Before I proceed to relate some of the unusual experiences throughout my career, I would like to explain how product design fifty years ago differed form today's approach, and how my personal design philosophy differed from the norm.

When I started my design business in the early nineteen fifties, one could almost say that industrial design was virtually unknown in Canada, partly because so few products originated here. It was a hard sell, to say the least – difficult to get commissions. Once I got a few products on the market, particularly as resident designer at Massey Harris (later Massey Ferguson), advancement was achieved through repeats and referrals. It really took about ten or twelve years to become established.

In the United States I am sure it was different. Raymond Loewy, considered the grandfather of industrial design, for instance, had been in the industrial design field since 1930. The Coldspot refrigerator he designed for Sears Roebuck in 1934 was one of his first major successes. However I venture to say that the postwar 1953 Studebaker Starliner Coupe (see photo),

acclaimed "the top car of all time" by Automotive News, with its innovative concave sculptured side panels, afforded him first notice with most Canadians. In 1957 his Sheerline ranges and refrigerators, with their sharp corners and flat panels, were well received in both the U.S and Canada. In fact this squared design has remained in vogue until the present.

When referring to industrial design, or product design, the term "styling" has often been used, even in reference to such as Raymond Loewy. This term is not in my design vocabulary. I suppose one reason for this is my exposure to the Bauhaus philosophy where function is paramount. This influence lingered on throughout my whole career, and I make no apologies. Though appearance has been a major part of industrial design ever since its inception, it must not be allowed to supercede the performance of a product. A chair, no matter how aesthetically pleasing, having a hard, flat seat and back support, is not comfortable and, therefore, does not constitute an acceptable design. For me the same emphasis on function applies to all well designed products.

These comments remind me to say a few words about the postmodern movement. In the 1970s there was a rebellion of sorts against the traditional design approach. Some have referred to it as anti-design, and I can see why. The main emphasis seemed to be on shock value, and disregard of logic. The most glaring example of this extremism was the Memphis group. The reference above to the uncomfortable chair applies here precisely.

Many, if not most product designers, when initiating the design process, will present a number of freehand sketches to illustrate various approaches, from which the client can then choose his preference, for further execution. This was not my approach. Being the designer, commissioned by the client, I felt that it was my responsibility to analyze all aspects of the assigned problem sufficiently to put together a well thought out best effort, upon which to base a final detailed design. Sometimes there were minor variations, as part of this first presentation, though it was a single unique concept.

The approach, then, was to start with a centre line, or datum of some kind, on the drafting board (no computers in those days), and produce an engineering layout. Since the typical client is not capable of judging a proposed design from an engineering drawing, it was necessary then to prepare a more pictorial presentation. Since the precise mechanical drawing now existed, rather than do a fee hand sketch I chose to do a technically accurate perspective rendering to show the client exactly how the product would look. Some of my renderings were actually mistaken for photographs (but the object did not yet exist).

In most cases, to illustrate a product to the best advantage the general case of perspective is required: three-point perspective, rather than two-point or one-point. In two- or one-point we are looking horizontally toward a horizon and all vertical lines are measurable; whereas, in three-point perspective there are three vanishing points, taper in all three directions and nothing is measurable. Instead of looking horizontally we are looking down at an angle at the product, or sometimes up, if the product or building is large or suspended overhead. The exception is with something large like a harvesting combine or an automobile where we would be gazing horizontally and would use two- or one-point perspective. What I would like to emphasize is that if it took an hour to manually lay out a two-point perspective drawing it could take a day to lay out a three-point, since it is so much more complex.

I will make no attempt to describe completely how to do such a layout. It would be boring, to say the least. I will just include the illustrations required when I explained a shortcut to producing a three-point perspective in the magazine "Product Design and Materials" back in December, 1957. And remember this was a shortcut (Plate 5).

Much later in my career I happened on a very simple way of converting a two-point perspective into a three-point perspective in a matter of seconds, saving a lot of time. I would lay out a two-point perspective using a vertical picture plane and a horizon, when the article should have been viewed at a downward angle .The resulting drawing would sometimes embody extreme distortion because of the article's being so far below the horizon (I am sure we have all seen photographs or drawings of tall buildings where the upper corner of the building appears to be less than 90 degrees, simply because someone thought that the building sides should appear parallel, and the picture was taken horizontally in spite of the building's height).In spite of the distortion I would complete the skewed rendering; then

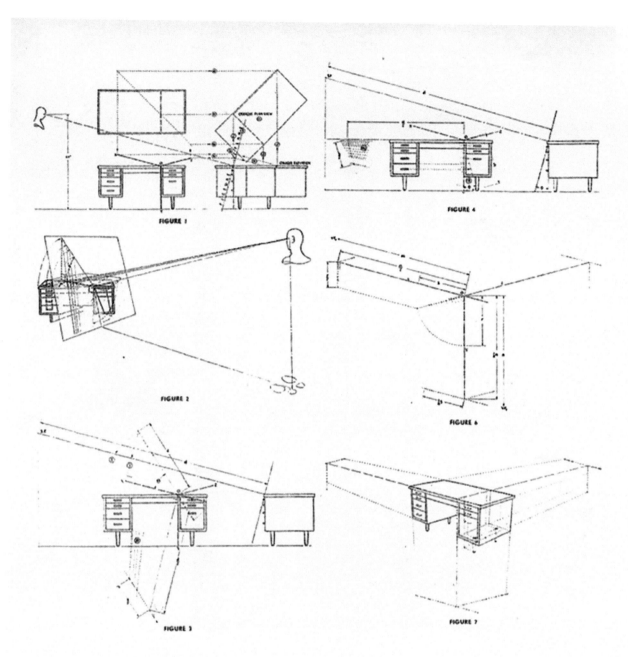

FIGURE 1

FIGURE 2

FIGURE 3

FIGURE 4

FIGURE 6

FIGURE 7

Plate 5.

Plate 5

I would place the camera at the station point in relation to the completed rendering, knowing exactly where it was, and photograph it at a downward angle, (Plate 6). The result was a perfect three-point perspective. The client would never see the distorted rendering but only the "corrected" photo of it.

Having gone through all this laborious effort for so many years I can only imagine how useful the computer is in solids modeling, especially since a rendering is no longer static but can be rotated for viewing from different angles. This is aside from the comparative ease of constructing the CAD drawings in the first place. I never had the advantage that a computer affords and finished my career working manually. When first introduced it would have cost close to $100,000 to adopt CAD. I kept postponing the innovation for years as the cost went down, and retired at 70, in 1994. At that juncture companies were beginning to ask specifically for CAD drawings, but that was not my reason for retiring.

While we are dwelling on the subject of the writer's distinct approach to the design process, we shall look at an event that occurred in 1976 that demonstrated a methodological conflict that arose on that occasion. We were commissioned by a Canadian manufacturer to design a new gasoline powered chainsaw. We took the assignment seriously and planned to help alleviate some of the inherent problems attributed to chainsaws. For the purpose a small team was assembled, including experts in internal combustion, noise abatement and vibration. Chainsaws had been notorious for excessive noise and vibration. Operators using chainsaws for extended periods were suffering hearing damage and "white finger" disease. Our intention was to reduce noise levels with a specially designed muffler system and reduce vibration by modified crankshaft design or vibration isolation mounts.

The first phase of the assignment, involving a five figure fee, would be consumed by noise and vibration studies and proposals; and the second phase, involving a six figure fee, would consist of testing developed concepts and integrating them into an assembled prototype that delivered a good performance. Phases 1 and 2 were approved for execution. The last phase would involve design refinement and production considerations. We had progressed part way through phase 1, according to our plan, when the U.S. parent company began requesting progress reports and visual presentations (sketches) to illustrate what this new chainsaw might look like. Progress reports should well be expected, but we were in no position, at this stage, to present sketches of something that had not yet been configured. It seemed that there was no way of convincing them to accept our approach. They cancelled the contract.

This was a case of disagreement on the whole design approach, which had been explained in our design proposal. I sued the company for breach of contract. Rather than sue for the total amount of phases 1 and 2, as I should have done, I sued for the modest amount required to fulfill my obligations to the personnel I had engaged for the project.

Pointing out the fact that there were professional engineers on both sides of the conflict, the judge recommended settlement. That was the way the dispute was ended. This was the only time during my career that my design approach was questioned, overtly at any rate, and so it must have had some validity.

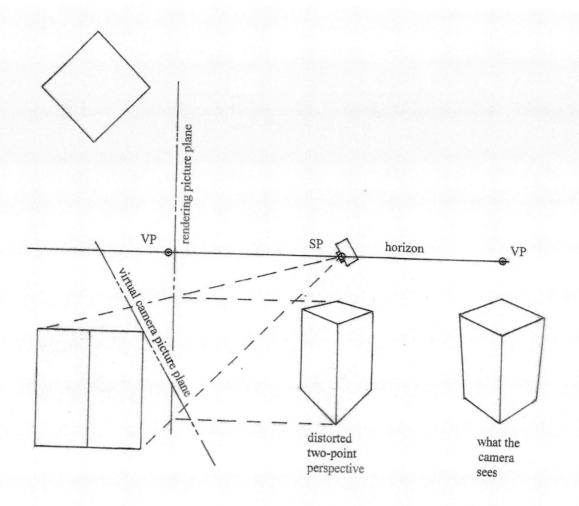

rendering picture plane

virtual camera picture plane

VP · SP horizon VP

distorted
two-point
perspective

what the
camera
sees

Plate 6.

Plate 6

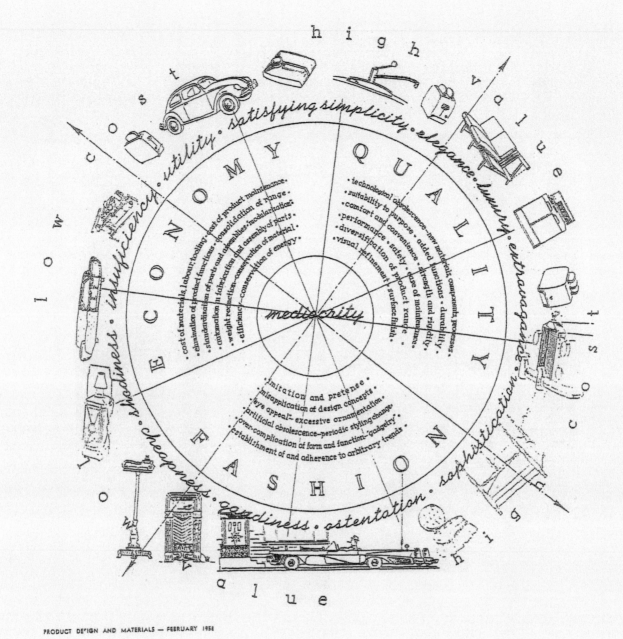

Plate 7.

Plate 7

CHAPTER 3 THE POLAR DESIGN GRAPH

The following is a two part treatise on the Polar Design Graph, as conceived in 1958, and published in Product Design and Materials. Much has changed in almost fifty years, as will be evident from some of the references. This should be borne in mind while reading the text and studying the Graph.

PART 1

What constitutes good design, bad design, — mediocre design? Is there a scale of values ranging in direct progression from good to bad, or is the picture more complex? How can specific design solutions be evaluated and apprasied? Would it be possible to devise a satisfactory system to assist in the orientation of design thinking?

It was in direct answer to these questions that the "Polar Design Graph" was conceived and developed. It is an attempt to present in graphic terms a comprehensive statement of basic design principles. Although it is a personal solution it seems to posses a degree of objectiveness in application. The following is a brief description and explanation of the graph.

Three Facets of Design

The three segments of the circle designated "Economy, Quality, and Fashion", represent the three basic facets of design manifestation. Contained within each of these segments is a list, by no means exhaustive, of typical design criteria — standards of judgement relating to the three facets respectively. It is the application, either systematic or intuitive, of such standards that determines the characteristics of the final product. These characteristics are set forth, in the extreme, by the descriptive words extending around the immediate circumference of the circle. In the centre is the middle ground of mediocrity.

The whole could be likened to a colour circle embodying three primary colours and their various intermediate hues, with distance from the centre denoting colour intensity and middle grey being central. The colour analogy extends even to the exsistence of a complementary relationship between any two mutually opposite positions on the circumference of the circle. Crosschecking of complementary meanings seems to verify the fundamental concept involved and suggests universal applicability.

The illustrative sketches around the circle are intended only to help clarify any ambiguity pertaining to the verbal descriptions, rather than to provoke extended controversy. Pictorial illustrations have their limitations in that many of the attributes of products within the various categories are a matter of comfort, convenience, feel, — values which can be determined only after prolonged personal experience with the kinds of products in question, and not by simply looking at them.

Arbitrary Examples

As result of the partial lack of such experience certain examples may appear to some extent arbitrary. A product that seems redundant in visible detail may be very efficient in operation, and a product having a neat and satisfying appearance may be lacking in an engineering sense. This contradictory aspect of the design of components makes it difficult to pass an opinion on the total design value as an integrated product. If asked to place any given product within the framework of the Polar Design Graph most of us would likely disagree to some extent. However if the intended concept is clear, that is all that matters.

As well as being divided into three equal segments the graph is bisected by two mutually perpendicular axes. One of these axes realtes to cost, and the other to value. The term "Cost" refers to the total cost picture as related to launching the product, cost of production, and cost of maintenance after purchase.

The term "Value" (high and low) was chosen rather than "Design" or "Taste" (good and bad) to establish our basic concept on as firm a premise as possible. Although value is certainly relative and becomes highly subjective in reference to asthetic qualities especially, it is much less subject to dispute and controversy than is "taste". Value here refers to total value represented by each dollar of cost inherent in the design. The most important revelation is that "good design" is not necessarily either expensive or inexpensive. It is available at any cost ranging from the very low to the high, and is represented by efficient and proper use of materials and processes available to satisfy existing needs.

Artificial Inflation
Our reference to cost rather than price is intentional. Market price is not always a reflection of production cost. Aside from the necessary absorbtion of costly advertising and merchandising expense, one of the most flagrant violators of the natural relationship between cost and market price is phenomenon known as "snob appeal". This is manifested as a combination of good value and novelty — the avant garde of design. The result is an artificial inflation of price because of the prestige value associated with "artiness".

Unfortunately, many well conceived products are subject, on introduction, to this influence. Indeed it seems to be the only way for some products to achieve eventual accpetance in the mass market at a reasonable price: a premiere engagement in the exclusive theatre of snobbery.

The influence of prestige on buying habits is important. Most of us have an unconscious tendency toward ostentation of some form or other. The whole field of consumer motivation is important in the choice of areas in which to concentrate during the process of planning and designing new products. The Polar Design Graph is intended to assist in the orientation of general design thinking and to suggest areas of investigation that might otherwise be neglected in the specific case.

PART 2
The Polar Design Graph is not intended to provide a ready made solution or even approach to any specific design problem. Its purpose is to help clarify in the mind of the reader the realtionship between the approach to design problems, their solutions, and consumer attitudes associated with these solutions.

One consumer is looking for the *least expensive* product that will satisfy his basic needs. Another is looking for the ultimate in comfort and convenience *at any price*. They may both be looking for "good design". Still another shopper may be looking simply for a means to satisfy his need for social recognition with more concern for outward display than for basic enjoyment and satisfaction.

Design Elements
As explained in our previous description of the graph, the descriptive words around the circumference represent extremes in product characteristics. Most existing products would actually lie closer to the centre because of conflicting attributes. For instance a well conceived and visually satisfying product might be the victim of shoddy internal detail design and construction which would decrease its value

to the owner. On the other hand another equally well conceived product of simple design might be spoiled by the addition of excessive and expensive surface decoration, which would likewise decrease its dollar value. In either case the resulting product tends toward mediocrity.

With these thoughts in mind let us examine the design picture with the help of the graph. Starting at the point of extreme economy at the upper left, we shall proceed in a clockwise direction making informal comments as we go.

The result of imposing the most stringent of economic limitations on the design process is usually an inadequate product. Not only are unnecessary functions and features eliminated to lower cost, but certain functions generally accpeted as essential are either omitted entirely or sacrificed in part by flimsy construction. The example given depicts an inexpensive combination shop tool designed to perform various functions, but none of them adequately. The saw table is too small, the bearings are undersize, the motor is underpowered, and the structure deflects under load. There are many other examples of this false economy, such as the stapler that jams or distorts the staple without clinching it properly, or the can opener that ceases to puncture lids after a few months' service. These products usually result from carrying manufacturing economy to the extreme.

As we move toward the top of the circle on the graph we pass through the area of "simple adequacy for a price" — the shopping centre for the thrifty consumer without a yen for prestige. Only the necessary functions are included, others being eliminated in the interest of economy. We have illustrated the now classic example amongst small cars.

Economy and Quality
Next is the area which taxes the designer most severely, the perfect balance between economy and quailty. Here we have the simplest solution that will provide complete satisfaction — usually the most difficult to achieve. A typical example of simplicity, though not lacking in quality, is the single control faucet to control both quantity and temperature of water delivered. Good examples in this category are all too rare. They are of the type that lead one to say on occasion, "Why didn't I think of that?" Ironically these are the problems that often require many hours of concentrated effort for the designer to arrive at the moment of insight which reveals the simple solution.

Proceeding around the circle we find the term "elegance". This refers to items with that extra touch of refinement that appeals to the discriminating buyer. The design is good mechanically, structurally, and visually, and nothing has been spared in employing suitable materials and achieving a fine finish. To be eligible for this category a piece of furniture, for instance, would possess a consistently good finish even where not normally visible.

These products constitute the avant garde and unfortuneately have a tendency to slip around the circle in a clockwise direction, as far as price is concerned, propelled by snob appeal. This was mentioned in our previous description of the graph.

Beyond elegance is "luxury", then "extravagance". Here we go past the simple and refined approach. we begin to add functions and features which may not be necessary but are desirable for those who can afford them. Typical of this category are the automatic timer on a range, the convertible car with the extra hardtop for winter driving, or the adjustable suspension on the Rolls Royce for rough or smooth road conditions.

Artificial Refinement

As we approach the point of balance between the quality and fashion influences, there is an area of artifical refinement exemplified by attempts to copy tradition — to produce, with today's high labour costs, the outmoded designs of yesteryear. Typical are period furniture and ornate lighting fixtures. These products are greatly influenced by fashion, in this case "old fashion".

As this influence, which we call fashion, increases our next area is almost completely at the mercy of the prestige demand. Products falling within this group are not only designed expensively but must *look expensive* as well. There is still an admirable degree of quality but the lavish and meaningless surface ornamentation accounts for much of the inherent cost.

As someone once said in speaking of the modern car: If they would only take all that chrome and sink it into a good non-corrosive exhaust system they would really be doing something. We heard the other day of a man who purchased an expensive car, and when the exhaust system failed a few months later, the parts alone, including four mufflers, cost One Hundred dollars. There is also a lot of "gadgetry" — merchandising features of dubious value which add unnecessary complication to products that could be simple.

Styled Obsolescence

As we go farther around the circle products begin to lose their quality, and sound engineering design gives way to the expensively contoured and decorated facade. This typifies the fashion influence. Products are "restyled" periodically to achieve an artifical obsolescence which we might call "styled obsolescence". This has become an important factor in our present day economy.

As we move around to complete our circle, and products lose even more of their intrinsic value, we encounter another kind of artificial obsolescence. It is inconceivable that anyone would purposefully search for a product of cheap or shoddy construction, regardless of apparent savings. Such purchases are usually the result of a combination of economic necessity and misleading outward appearance or advertising claims.

A few years ago we bought a chesterfield for our own home, not unlike that illustrated at the extreme left side of the graph. The trap was nicely set, with the purchase being made under the very circumstances mentioned above. Our budget was limited, the product looked pleasing enough, and the manufacturer was "name brand". The price paid was $150. When the chesterfield began to disintegrate three or four years later we were more annoyed than surprised to discover that, despite its cumbersome bulk and weight, it was fabricated in large part from sheets of cardboard and fourth grade lumber, cheap stuffing and covered with a light fabric susceptile to wear and fading.

We vowed then and there *never again to buy that brand of furniture.* We replaced the discarded item with one like that illustrated at the upper right side of the graph. The price was close to $300 dollars. It has an extremely rigid structure, with only a fraction the weight of the piece it replaced, it is pleasant to live with, and above all it offers a new experience in comfort for anyone who has used it in our home. It will last until we get tired of it, and that will be a long time.

We have learned our lesson, as many other consumers have, but has the manufacturer learned his lesson? The manufacture and sale of flimsy products with the express (if not expressed) purpose of maintaining a recurrent market demand, may backfire for the manufacturer concerned if pushed too far. Customers may be irretriveably lost, and the economic argument of increased productivity and higher living standard may break down. It seems evident that by buying a product of lasting quaility the consumer is then freed from the necessity of a repeated purchase of the same kind of item, with a facelift, within a comparatively short interval of time. Instead he can raise his living standard tangibly by turning his attention to other products, such as freezing or air conditioning equipment, the results of new technology and creative effort.

What do consumers want?
The usual defensive argument of the manufacturer is that he is simply giving the consumer what he wants. I am sure that no one really wants shoddy design and construction (which is by no means limited to inexpensive goods), and he will not knowingly buy it a second time. He may want something different. As stated before, the designer is at his best when evolving solutions that are completely satisfying but at the same time economical.

Consumer surveys may reveal that certain items are replaced according to a definite periodic pattern. What they may not reveal is that this replacement is not necessarily voluntary. We have the suspicion that many consumers spend their shopping hours desperately trying to find products that will last. Each time they buy a different brand, not because they are fickle but because with each purchase they hope that they may step off the merry-go-round for a while to gain equilibrium. As a result of conditioning they finally accept the transitory nature of our present day economy and give up the search.

Conclusion
This leads us to our conclusion. An examination of the market of available products reveals a definite overemphasis on wasteful extravagance. This is not limited to luxury items but extends all down the line to products that are intended for the "low price" field. For instance the "small" cars are now bigger than the big cars a few years ago, but they still carry only six passengers. Whether this general trend is a result of a genuine demand or the conditioning of costly merchandishing is a debatable question. There is evidence of an open rebellion amongst consumers. To save the situation, some concentrated design effort is required that will shift the centre of gravity back toward a better dollar value.

CHAPTER 4 ANCEDOTAL EVIDENCE

The many design projects in which I have been involved fall into three main categories.

1. Projects involving industrial design in its accepted sense of providing exterior design and refinement while satisfying ergonomic requirements. Examples would include electronic packaging, whether an intrusion alarm, a telephone exchange, an electron microscope or a defibrillator. They would also include cancer therapy machines and harvesting combines. Typically the company would have a complete staff of engineers and scientists.

2. Projects involving both engineering design and industrial design, usually smaller, less complex products. Examples would include electric kettles, toasters, domestic and industrial wiring devices, desk calendars, hair dryers and gas barbeques. Client firms would usually be smaller, have minimal or no technical staff and be more dependent on the design consultant. Working drawings would be part of the job .

3. Projects involving engineering only, with less concern for the final integrated form of the product. Examples would include truck loading dock ramps and production line machines.

Before presenting any specific design anecdotes I would like to include here some advice to students or designers in the field, or design firms, for that matter. I found that it was always better to estimate, rather than quote on projects. It allows for some variation, and not only on the plus side. I remember that on at least one occasion the job turned out to be less onerous than I expected, and I charged the client considerably less than the estimate - the result, one happy client. Also I tried to account for contingencies when estimating. I made it a point to prepare as accurate a time summary as possible, and then submit a proposal covering about twice that time. If the estimated time of delivery was six weeks and I delivered in four – another happy client. That's important. As an independent designer I always separated my office time from my family time and seldom worked evenings or weekends. It's simply a matter of discipline and time allocation. Never promise what you can't deliver. Some other designers were slaving in the wee hours for presentations the next day. Of course, I realize that for employed designers such decisions may be entirely out of their hands and they may not be able to implement many of these suggestions.

Most, but not all, of our design efforts were expended on products for quantity production. It might be an eight-track tape cartridge produced in quantities of eight million per year, a hair dryer or computer keyboard tray in total quantities of half a million, or a cancer therapy machine or telephone exchange in quantities of ten or twenty per year, but costing hundreds of thousands of dollars each. Occasionally there would be a one-off production line machine, an engineering job.

We will now recount some random anecdotes and observations concerning our activities throughout the years, and hope that they engender some interest.

VANITY OF VANITIES

One of the first clients was an electric and gas range manufacturer in Montreal. This was a category 1. project, and consisted of exterior design with some thought devoted to ergonomics, especially where knobs and controls were involved. Though it was slightly before my time, one interesting "ergonomic" requirement had been that all ranges sold in Quebec required a mirror on the backsplash – only in Quebec, you say!

TOXICITY

Another product from the early fifties was a steam iron, Page (41). People have asked why the iron was not designed to stand on its heel as most irons are today, so that its sole plate would not be in danger of overheating or burning the ironing board surface. The simple answer is that we designed a framed pad for the iron to rest on – an asbestos pad, another forbidden substance! Again we didn't know that asbestos presented a problem.

Speaking of forbidden materials, when I was working on the design of sheet metal parts, away back then, someone suggested manipulating a soft, pliable metal to simulate formed steel parts. In this way I could establish the shape of the steel blank required to form a finished part. The material was lead. Anyway I am still here.

JUST TESTING

Back to the fifties. One might say that I was involved with "everything but the kitchen sink". Well I did design a kitchen sink and also a toilet, the kind that is used in a cottage, and requires very little water to flush. While I was working on the project, my client mentioned how he tested the toilets. I thought it was amusing, but when I laughed he was upset. He used peeled ripe bananas!

MUSTANG

The tractor prototype, designed in 1954, and shown on page (45), was never produced and sold (a concept tractor ?). However, I believe I am correct in saying that the Ford Motor Company sought clearance from Massey Ferguson before using the name for its new car introduced in 1964, apparently the most successful product launch in automotive history.

AVOIRDUPOIS

The Theratron 60, mentioned on Page (46), was the first cancer therapy machine that I worked on, in 1959. The machine that it succeeded had embodied a certain amount of tungsten for shielding around the cobalt 60 source (a capsule of cobalt pellets), the remainder of the shielding being lead. Tungsten was a better shield but was also much more expensive. In order to reduce cost, this tungsten was replaced by lead. For adequate shielding eleven inches of lead was required. This means that there was a twenty-two inch diameter sphere of lead inside the head. A quick calculation shows that this amounts to more than 5,575 cubic inches, at 0.4 lbs. per cubic inch. The lead sphere alone then weighed 2,230 lbs., and with the cast steel enclosure the total head weight would exceed that of a small automobile. Because of the balancing counterweight and the remaining cast steel structure, the whole machine weighed several tons, I believe seven.

FLOOR SPACE

A few years later, in the sixties, I was engaged by a firm in North Carolina to design some nursery furniture, a table and two chairs, see photo. These were successfully marketed, but only after an unexpected price adjustment. The tooling cost for injection moulding was very high, but the part costs were very low – so low that the proposed selling price on the sales floor did not warrant the floor space required. The solution, of course, was to raise the price, and apparently it worked.

REDUNDANCY

One of the engineering jobs that we were assigned was to redesign an existing manually operated truck loading dock ramp, with the main goal being cost reduction. Referring now to Plate 8, the patent drawings show the general configuration of the ramp. There were two helically wound torsion springs 25, side by side, both of which were in a relaxed position when the arms 20 were vertical as in Fig. 5. As the ramp was eased backward, one of the springs twisted against a stop, counterbalancing the load and lowering the ramp gently into the stored position, Fig. 4. As the ramp was urged forward from the Fig.

5 position onto the truck bed, the other spring twisted against its stop and eased the ramp into that position. The ramp was operated by the use of the manual lever 33.

It occurred to me that since there were two identical springs, each being used half of the time, there must be some redundancy. The result was that we eliminated one spring and introduced a stop arrangement at each end of the remaining spring so that it was wound from one end going forward onto the truck bed and from the opposite end going backward to the stored position, Fig. 4. For anyone interested, a brief examination of Plate 9, Fig. 2 will reveal the method. Elimination of redundancy is an easy way to reduce cost.

With respect to this same product, it was proposed by the engineers from the steel supplier that further cost reduction could be implemented by switching to a new "high strength" steel, recently introduced. Since this steel was apparently 25% stronger they suggested that the 5/16 in. thick checker plate used for the ramps could be reduced to ¼ in., with substantial material savings. I suggested to my client that perhaps they were mistaken, and so I did some calculations. Since the strength of a plate varies not as the thickness, but as the square of the thickness, reducing the thickness (of the steel presently used) to ¼ in. would reduce the strength to 64%. Now if the high strength steel is substituted that would increase the strength by 25% back up to 80% of the strength of the original 5/16 in. plate – not good enough. And furthermore the stiffness of a plate varies as the cube of the thickness. Then the ¼ in. plate would lose 49% of its stiffness.

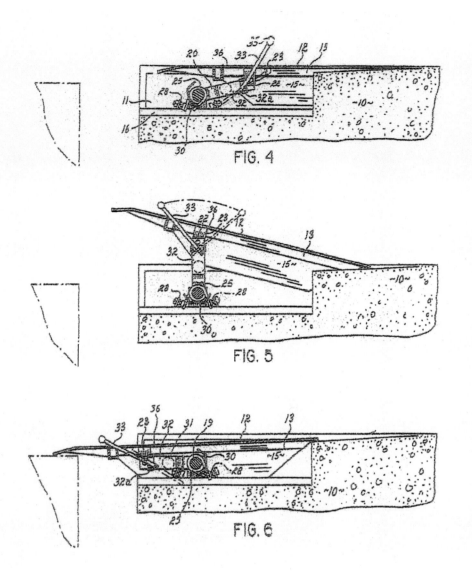

FIG. 4

FIG. 5

FIG. 6

INVENTORS

Lawrie G. McIntosh
John A. Merrick

Agents for the
Applicant

Plate 8

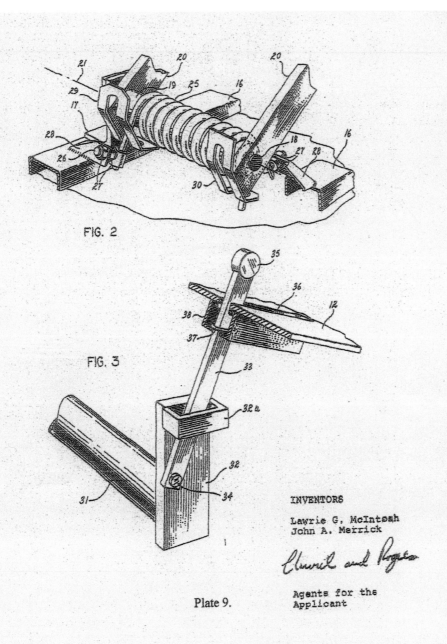

FIG. 2

FIG. 3

INVENTORS

Lawrie G. McIntosh
John A. Merrick

Church and Rogers

Agents for the
Applicant

Plate 9.

Plate 9

Unfortunately my client put his trust in the steel company engineers and followed their advice. After all there were two of them, and I was only a design engineer. He switched to the thinner plate. A few months later caved dock ramps were being returned. From then on my client trusted me.

SYNCHRONY

Perhaps the most difficult engineering undertaking we grappled with was an assignment from a large oil company to insert small plastic playing cards into cans of motor oil. The idea was to encourage service station attendants to ask customers if they wanted their oil checked. If they found a playing card in the can after they dispensed the oil there could be some financial benefit for the attendants. The extra challenge was that the company was already inserting plastic coins into the cans, but they wanted to change to the playing cards and double the filling rate to ten cans per second. One card was to be inserted into every fourth can.

Initially it seemed insurmountable, especially since a card might flutter and miss the can. Everything had to be synchronized perfectly. First we obtained permission from the production people to drive our machine directly off the production line, giving automatic synchrony. That helped. Then we designed a mechanism to first pluck a card, using a suction cup, from a vertical stack of cards in a magazine; and then in a sweeping path, to deliver the card to the proper can at synchronous horizontal speed. The final refinement was to deliver the card with a puff of air through the suction cup. All of this mechanical action was repeated every 0.4 seconds. What still amazes me is that they could fill ten cans every second.

LONGEVITY

This is just a comment about a couple of products. The electric kettle on Page (40), the one with the combined handle and spout, has been on the market, unchanged, for 36 years. For a small appliance this may be some kind of record. Also the computer keyboard tray, Page (43), remained on the market in various versions for about seventeen years. Since computer models themselves might last only two or three years, this too seems a rather long market exposure for ancillary equipment.

ALMOST BUT NOT QUITE

While I was working with another office furniture components company, particularly on chair mechanisms, we turned our attention to gas springs used for height adjustment. There had been reports of serious, even fatal, accidents involving these pneumatic mechanisms. They contain gas under pressures of two to three thousand pounds per square inch. If someone tinkers and causes an explosion it can have serious consequences. I began working on a mechanical version that would accomplish the same action as the gas spring, with some success.

The same kind of gas spring is used to support hatchback and trunk lids on automobiles. Because of the millions of cars requiring these springs, and the hazard involved, the company agreed that I should devote some attention to this application (I dreamed of royalties). We actually succeeded in designing a mechanical version, using square section wire spring to pack maximum energy, and presented it to General Motors in Detroit. They were sufficiently interested to have the Fisher Body division fabricate ten prototypes for testing. On testing they were found to perform quite satisfactorily. Things looked great. However, when subjected to 250 degrees F., they lost 15% of their energy and failed. Such temperatures are possible inside a car trunk.

In the latter part of my career I was involved in patent and industrial design litigation as an expert witness in court. Some of these cases were boring and some were interesting. We will look at some of the latter.

MINE VENTILATION

There was a firm producing ducts for ventilating mines. An employee of the firm decided to leave and start his own company, producing the same product. The original firm was not concerned about his producing the same product but suspected that he had copied their production machine. The firm lawyer engaged me as their expert witness to investigate.

I was asked to visit both manufacturing facilities and photograph both machines, for comparison. The machines appeared to be identical. My approach was not to compare the process involved in production but to concentrate on the visual or industrial design aspects. I concentrated on the base of the machines. In court I indicated that the least arbitrary design for a base would be a round (or maybe square) shape directly under the centre of gravity to provide equal stability in all directions. However the base on the machine in question was attached to the bottom of a vertical pillar and extended out as a rectangle with a taper at the outer end – the exact same shape as the base on the original machine. This was enough to convince me that both machines had been fabricated from the same drawings, drawings taken from the files of the first company. The judge agreed, and so industrial design matters saved the day.

COFFEEMAKER

A Canadian entrepreneur commissioned a Montreal designer to design an electric drip coffeemaker. He had it tooled in the New Territories, adjacent to Hong Kong. Then he sold the design to a major American small appliance company, and had the product made in Hong Kong. After a time the American firm decided to buy direct and exclude the entrepreneur. The entrepreneur then sued the American firm and engaged me as his expert witness.

In court, we displayed one of the coffeemakers in question, and the opposition lawyers brought in one from their production and about half-a-dozen other drip coffeemakers of various brands. Their expert began expounding on the idea that all coffeemakers had a carafe, a chamber for water, a basket for coffee grounds, a tube for conducting the water to the basket, a cord, etc. The implication was that all coffeemakers are not only similar but virtually the same. I countered with: since a VW beetle and a Rolls Royce both have wheels, a motor, seating for five, a steering wheel, etc., they are similar and virtually the same. The judge took the point, but what really was the clincher was something else that I discovered.

On examining our coffeemaker and comparing it with the one that they had brought in, from their production, I found some fine marks on their sample that exactly matched marks on our sample. These turned out to be scratch marks on the mould reflected on the parts. This proved that not only did they copy the design exactly but they used the entrepreneur's moulds for early production, before they made their own moulds. The judge was convinced.

Having won the case, the entrepreneur next sued the manufacturer in Hong Kong. Quite a few months had passed before the case came up in Hong Kong, but I was asked to attend again there as expert witness. According to Hong Kong law, at that time at least, an expert witness was not permitted in the court room except to give testimony (not the case in Canada). Since my previous testimony was common knowledge to both sides, I was never called to the stand, but, of course, my presence was all that was necessary to trigger a substantial out-of-court settlement.

For me, except for two or three days hanging around the court, it amounted to a paid vacation since I had a few days left before my flight departure. I managed a walking tour of Hong Kong and Lantau Island and a jet-foil trip to Macau, before returning home. Also I received $5,000 in fees for just being there. The entrepreneur was obliging and generally pleased.

SHOWER HEAD

In another case, before I went on the stand, the opposing lawyers were discussing with the judge a possible precedent setting case in which he had been involved. They asked him if he remembered how the particular case ended. He said that no, when he left the court he tried to relax and forget what happened. The next day I was on the stand experiencing the usual attempts by the opposition to discredit me as an expert witness before cross examination. They knew that I had worked before with the same lawyer and were trying to capitalize on it. Anyway they brought up a previous case where we had collaborated and asked me if I remembered how it turned out. I really didn't remember, so I said: "No, when I leave the court I try to relax and forget"----. The judge said: "Well you're on safe ground there". Everyone laughed and the lawyer immediately ceased his harassment.

ACADEME

I should mention that, along with my professional design work, I lectured at various universities and colleges in Ontario. It began as a fill in when I was not busy early in my career, but soon my contributions were in demand and I eventually achieved an "emeritus" standing at the Ontario College of Art and Design. In particular, I taught a class per week there for extended periods of time between the fifties and the eighties. My main purpose was to impart basic engineering principles to the industrial design students. I had always felt that many industrial designers tend to make design decisions based on intuition alone, not realizing that what seems to make sense may not be logical after all.

To illustrate the point, I refer back to page (26) where it was shown that variation in the thickness of a sheet of material produces an effect that may defy intuition .This is just an example of how design must be based not only on intuition but on knowledge of strict learned scientific principles that cannot be challenged. Some of the other topics for study and discussion were tension and compression, vectors, resolution of forces, structure, weight and mass, dynamics, etc. It was difficult, at first, to communicate these concepts to my students since they were, after all, in an art college; but I know that they appreciated their application of my instruction as long as I avoided involved mathematics.

I have found that of my two academic fields of study, mechanical engineering and industrial design, the former contributed as much as, if not more than the latter to product design problem solving. For students contemplating a career stressing genuine product design, as opposed to the visual arts such as packaging and corporate identity, I would strongly advise considering an engineering degree to augment other design training.

Back to the Ontario College of Art and Design, to supplement classroom instruction we made field trips to manufacturing plants in the area, some of which were those of Frank Stronach's Magna International. In these factories there was a great deal of automation, for instance the stamping out of complete van side panels in one piece, front to rear. This was accomplished in successive "hits" inside a glassed–in tunnel, with all personnel excluded before production could begin.

On one occasion we visited a Magna metal plating plant. Plating plants at that time had been notorious for emitting very toxic waste. However, while the students and I were in the plant reception area we noticed a big glass tank, or aquarium, with rather large fish actively swimming about. I was curious about the tank and was told that the fish were swimming in the plant effluent. Years later I read a woman's letter to the editor in a local paper criticizing firms like Magna International for having no regard for the environment and, therefore, causing extensive damage throughout southern Ontario. Little did she know! Too many people make too many unsubstantiated assumptions.

CHAPTER 5 RETIREMENT

After a fulfilling career I retired at 70, in 1994. Since my work had been like a hobby to me I had to find something else interesting to do; and so I traveled extensively for a few years and prepared several travelogues for the record. They included: Trekking in Nepal; U.S. Canyons; San Francisco and Napa Valley; Greece, Turkey and the Greek Islands; The Canadian West Coast; Fiji and Australia; and the Atlantic Provinces. In my travels I became intrigued with certain business signs and began photographing them. Puns had always intrigued me and all of these signs embodied puns, most of them intentional, but some unintentional – some of the foreign ones. I have included two photomontages (Plates 10 and 11), a small sample of the many photos taken.

About eleven years after my retirement, the Association of Chartered Industrial Designers of Ontario (ACIDO), and the Design Exchange (DX), together, staged a retrospective of my career accomplishments and mounted an exhibition at the Design Exchange in Toronto, from January 27th until February 18th, 2005. It was called "Impact 2" and consisted of a number of three foot square panel boards whose content is shown in the following pages. Because I have shown all of the board content, there is some repetition. I considered this show a great honour and a pleasurable experience, meeting old friends and new acquaintances.

Since the Canadian Animal series shown on page (50) in the retrospective is the early flocked vinyl version, I include a photo of the replacement copper version (Plate 12) at the end of the retrospective.

Plate 10

Plate 11

Impact

Industrial design has tremendous impact on the individual, society, the economy, culture and the environment. IMPACT, an annual design exhibition presented by the Association of Chartered Industrial Designers of Ontario (ACIDO), seeks to raise awareness of the work of great Canadian Industrial Designers.

LawrieMcIntosh

Lawrie G. McIntosh's distinguished career is notable for the wide variety of projects and industries to which he has applied his skills and knowledge. McIntosh, who was born in Clinton, Ontario on June 24, 1924, is responsible for significant innovation in the areas of medical and agricultural machinery, household appliances, office equipment and furniture.

McIntosh graduated from the University of Toronto in 1946 with a Bachelor of Science in Mechanical Engineering. In 1951 he obtained a Master of Science in Product Design at the Illinois Institute of Technology where he was taught by Buckminister Fuller. Throughout his career, he has been an active member of the design, engineering and manufacturing industries. McIntosh has participated in professional associations and taught at universities and colleges across the country. His articles on design have appeared in numerous trade publications, and he was Consulting Technical Editor for Product Design and Engineering Magazine.

A prodigious designer, McIntosh's awards include: the National Design Council Chair Competition in 1952, the Milan Triennale in 1956 (Gold Medal), the Stainless Steel Competition in 1961, and the Canadian Souvenir Competition in 1961. In 1981, he received a Citation from the National Design Council, Design Canada and the Association of Canadian Industrial Designers in recognition of his outstanding achievement and contribution to the practice of design and the profession of design in Canada.

McIntosh retired in 1994 and now lives in Calgary.

McIntosh Design Associates
Sample Product List

Consumer
Electric and Gas Ranges
8-Track Tape Cartridges
Room Air Conditioners
Forced Air and Convection Heaters and
Furnaces
Portable Typewriters
Refrigerators
Vacuum Cleaners
Electric Irons
Toasters
Hair Dryers
Food Mixers
Electric Kettles
Domestic Wiring Devices
Mops
Brushes
Nursery Furniture
Toys
Stationary Products

Industrial and Agricultural
Automatic Telephone Exchange Equipment
Burglar Alarms
Electronic Packaging

Cobalt 60 Irradiators and Controls
Lift Trucks
Truck Loading Rams
Industrial Wiring Devices
Harvesting Combines
Balers and other Farm Equipment
Live Stock Watering Devices
Centrifugal Pumps
Plumbing Fittings
Low Pressure Hydraulic Devices
Textile and other Process Machinery

Commercial and Institutional
Cobalt 60 Cancer Therapy Machines and
Controls
Linear Accelerators
Hospital Beds
Pulse Rate Meters
Medical, Dental and Lab Equipment
Electron Microscopes
Trailer Equipment
Beverage Vending Machines
"Enterphone" for Apartment Buildings
Washroom Odorizers
Desk Calendars

Impact: Lawrie McIntosh

Presented by

With Support of

 DESIGN EXCHANGE
MUSEUM · EDUCATION · RESEARCH

With Special Thanks to the Following

Impact Exhibition Concept, Design and Development
Tasneem Babul Rayani

Initial Art Direction
Sang Nguyen

Graphic Design and Layout
Tiffany Minorgan

Intro Board Design
Gottschalk+Ash International

Sponsorship Coordinator
Tim Poupore

Project Support
Lawrie McIntosh
Elise Hodson
Paul Arato
Randy Tarr
Scott Grant
Martin Clarke
Rona Arato
Farhez Rayani

Project Descriptions
Lawrie McIntosh
Elise Hodson
Tasneem Babul

References
McIntosh Design Associates Portfolio
Design Exchange Archives
Rachel Gotlieb and Cora Golden. Design in Canada Since
    1945: Fifty Years from Teakettles to Task Chairs . Toronto:
    Alfred A. Knopf Canada and Design Exchange, 2001.
Atlas Steel: Steel News (Vol. 19. Nos. 3)
Canadian Consulting Engineering (Dec. 1978)
Canadian Plastics (Nov. 1965)
Design Canada : Danfoss Tempress Automatic Shower Valve
    Case Study
Design Canada : Exerpacer Case Study
Design Canada : Atomic Energy of Canada, Ltd., Cobalt 60
    Teletherapy Unit
Product Design and Materials (Feb. 1958)
Product Design and Value Engineering (Aug. 1966)
Professional Engineer and Engineering Digest Product
Engineering (1960)

Design Philosophy

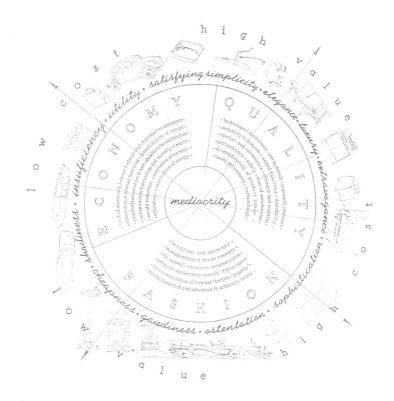

The first Canadian engineer to become an accredited Industrial Design in Canada, Lawrie McIntosh brings his analytical and creative skills and his intuition together to successfully solve problems. Even before finishing his post graduate studies, his work was gaining him the attention of renowned designers.

Always willing to teach and share his knowledge, McIntosh has contributed regularly to engineering and design publications and associations. First published in Product Design and Materials , McIntosh's Polar Design Graph (right) illustrates the basic design principles he applied in his practice.

McIntosh believes that to design a successful product, the intended function (be it mechanical, ergonomic, social, psychological or cultural) must be considered first. This is then followed by the structure and the form of the product – how it is manufactured, the materials used and the assembly required. The design process and the resulting product are gauged by the quality and the cost requirements of the product.

Illinois Institute of Technology

Fiber Board Chair, 1948

This organic looking chair was made by folding laminated sheets of fiber board together and was designed while McIntosh was a post-graduate student at the Illinois Institute of Technology. Hans Knoll was so impressed by the solution to this student assignment that he offered to hire McIntosh to open his new Canadian office. Despite being forced to decline such an amazing opportunity due to personal reasons, McIntosh was to embark on a successful career traversing many fields.

Lawrie

Steam Electric Products Ltd.

T-800 Op'nTop Kettle, 1962

This kettle was originally designed as an electric tea maker for Steam Electric Products, Ltd. with an insert for loose tea and a removable lid. The design was re-released as an electric kettle, emphasizing the open top. The design was later sold to Landers, Frary and Clark (Canada) Ltd. As model T-4-6.

McGraw-Edison later produced the kettle under the name of 'The Ultra'.

Landers, Frary & Clark Ltd.

Flat Top Kettle (The Ultra, Model 30430) , 1960

When the manufacturer approached McIntosh to restyle their electric kettle using nickel stainless steel, they challenged him to design a product that would stand out from the competition, yet not be radical or extreme. They also requested that the base circumference and volumetric capacity remain the same as their existing product line.

McIntosh decided to move away from the traditional hemispherical shape and responded with contemporary styling of large, gentle curves and crisp lines demarcating surfaces. By flattening the unused upper portion of the hemisphere above the spout level, the design improved balance, handling and structural strength. The nickel stainless steel allowed for a deep draw die and highly polished finish.

McGraw-Edison of Canada

Electric Kettle, Model 30380, 1969

Described in their catalogue as the 'top electric kettle value in Canada,' McGraw-Edison boasted that the two-quart kettle would boil in two minutes with guaranteed reliability. The innovative handle and spout gave the kettle a distinct advantage at a time when Canadians were buying 500,000 kettles annually (the world's highest per capita sales).

McIntosh replaced the traditional compression molded black phenolic handle and separate stamped and soldered spout (the soldering was expensive) with a single injection molded polypropylene piece. This allowed for a leak proof spout and handle assembly. The material was available in a variety of colours to match the new colourful appliances of the time. The kettle was offered in avocado, gold, white and black.Superior Electric Ltd. Of Pembroke, Ontario purchased the dies from McGraw-Edison and successfully continues to produce the kettle to this very day. It is also sold under the Toastess brand.

Winner of a National Industrial Design Council Design Index Award (1954) and a Gold Medal at the Milan Triennale (1956).

McGraw-Edison of Canada
Steam-and-Dry Iron, 1954
Designed with B.H. Pickard

This modern, streamlined design came in two models, the "Berkley" under the Eaton s Department Store brand name, and the "Vapor Jet R600." The fully automatic iron was known for its light weight and came with a plastic bottle to fill the iron with water.

The thermoset resin handle was influenced by an iron designed by Thomas Penrose for Canadian Westinghouse in 1952. The open handle allows the iron to go deeper into corners and cuffs.

Steam Travel Iron, 1953

Rotor Electric
Lady Torcan Hairdryer, 1960

One of McIntosh's favourite projects, this stylish hairdryer responded to a number of design problems. Previously, hairdryers had consisted of a simple box with rounded corners, a fan inside and a door behind which to store the hose. Typically, the door would pop open and the hose would dangle from the shelf where it was stored.

McIntosh created a circular housing that held a centrifugal fan in the centre around which the hose was securely wrapped. This allowed for an inexpensive and rapid tooling process using spun aluminum parts resulting in lower upstart costs for manufacturing. The product went from conception to market in only four months. The hairdryer proved extremely popular and the manufacturer was able to invest in more sophisticated tooling. Steel stampings eventually replaced the metal spinning and the unit cost was greatly reduced, as steel is a much less costly material. More than half a million units were sold in the 1960s with virtually no design changes.

Lawrie

Cords Canada

Various Applicance Plugs, 1950
Cube Tab, 1958
Trouble Light, 1959

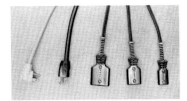

Various Appliance Plugs   , 1950

These removable kettle plugs, the largest shown (left), received a National Industrial Design Council award. Kettle cords to follow moved from this style to permanently attached cords.

Cube Tap,   1958

Prior three-way extension cord receptacles (or cube connectors) were much larger and had internal electrical contacts made from several pieces. When these plugs were inserted, their blades met at the centre (Fig. 1). Larry Caplan of Cords Canada approached McIntosh to help design a new method of reducing the size of the cube connector by reorienting the three receptacles and simplifying the metal contacts.

The existing contacts were made from two formed pieces of light brass riveted to a heavier strip which had a tapped hole for a screw (Fig. 2). The new contact is a one-piece contact crimped onto the wire with no screws and made on progressive dies that makes it very inexpensive to produce (Fig. 3).

The final configuration for the connector is derived by offsetting the plugs slightly so that the blades bypassed each other at the centre and still has enough room for the wire contact. (Fig. 4). The resulting form encapsulates this new configuration, creating a streamlined unit (Fig 5).

Fig. 1

Fig. 2

Fig. 3

Fig. 4

Fig. 5

Trouble Light   , 1959

This very recognizable work lamp which can be found in many a garage had its handle redesigned by McIntosh for Cords Canada. McIntosh designed indentations in the handle for a better grip and chose to make it out of PVC because of its durability and resistance to impact.

Lawrie McIntosh

Warwick Bros. & Rutter
Desk Calendar, 1961

Previous designs for desk calendar pads have hinged arches that are cumbersome and expensive. McIntosh's design, which was unfortunately not patented, features a spring wire arch that is widely copied in the U.S. and Europe.

Commodore Typewriter Co.
Portable Typewriters, 1960-1961

Series 1000 and 1500 won an Award of Merit from the National Industrial Design Council of Canada (1960).

McIntosh worked with the Accurate Mold Company to launch Commodore's reputation as a pioneer in the application of plastic to portable typewriters. The Series 1000 and 1500 were introduced in early spring 1960, just a few months before the American company Remington Rand developed their own. The award winning designs led to a portables exhibition in Canadian booths at European and U.S. trade shows.

The typewriter frame and base are integrated into easy to clean and assemble units that also form the lower portion of the carrying case. The base and cover, with indentations for the handle are made with a reinforced polyester.

In 1961, the Series 600 again challenged American competition with an injection molded Cycolac model. McIntosh designed a lightweight, compact, less expensive product, with a modern look that had proven popular in the first series. The matte finish did not show scratches, which appeared quickly on other shiny, smooth portables. Consumers preferred the 'quality' appearance of the subdued surface to the high gloss associated with 'cheap plastics

McIntosh

CompX
Adjustable Keyboard Tray, 1986

The main competitor, in the US, produced a version that had a horizontally sliding (for and aft) carriage for under the desk storage, and a vertically adjusting tray mounted to the carriage – two separate motions. Their design was covered by a firm patent and would have required a royalty payment to the US company. McIntosh's new design embodied only one arcing motion that provided both vertical adjustment and storage under the desk. This design circumvented the existing patent and produced its own patent. McIntosh received royalties until 2004.

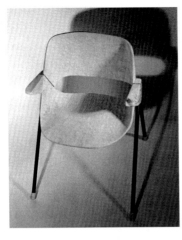

Aero Marine Industries
Plywood Chair, 1952

Made of plywood, solid wood and seamless steel tubing, this chair edged out 244 competitors to capture the $1000 first prize in the 2nd annual National Product Design Competition sponsored by the National Industrial Design Committee of Canada.

In addition to being comfortable, well-priced and easy to fabricate, the chair's knock down design makes it easy to ship flat, while also being easy to assemble for the end user. The judges representing the design, manufacturing, retailing, education and research sectors felt that the chair showed, "a thorough comprehension of the possibility and limitation of each material."

Province of Manitoba
Canadian Animal Series, 1962

Polar bear copper outline

1966 Winner of the Canadian Souvenir Competition
Sponsored by the Province of Manitoba for the Centennial, the original series was made from flocked vinyl, which then was replaced in the seventies by a more durable, copper series.

Each animal is made from a flat sheet of copper, where each animal's outlines are in perfect register on both sides. A resist is printed on each side of the copper, denoting the etch lines. Because the copper is severed at the outlines, when the sheet is etched from both sides at once, the animal template will fall out of the sheet. The fold lines, having been etched from one side only, are etched half way through. Concave fold lines are on one side and convex on the other. After folding and soldering, the finished animals are selectively tarnished, or in the case of the polar bear plated, for completion.

Massey Harris

(now Massey Ferguson)
Combine, 1952
Baler, 1953
Mustang Tractor Prototype, 1954

Mustang Tractor Prototype , 1954

McIntosh

Baler ,1953

In the original version, the container for the four balls of knotter twine was referred to as a "twine can" – an indication in itself that structural integration had never occurred to the engineers involved. Enclosing the twine within a compartment of the transfer chamber simplified structure, reduced cost and improved appearance.

Baler concept rendering

Combine ,1952

In the fifties, combines (left) had an open platform for the operator. There was no closed cab, let alone air conditioning, which came later. With this self propelled harvesting combine, McIntosh attempted to lower the silhouette to allow the machine to be stored inside a building, to provide some visual integration to the total assembly, and to allow easier servicing of the grain tank and other higher components (see before/after).

Combine concept rendering

before

after

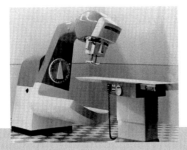
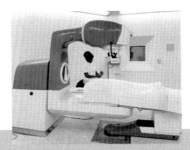

Theratron/Cobalt 60 Therapy Unit,   1959

Uses a capsule of cobalt 60 pellets, which emits gamma radiation, to reduce tumors. Because the cobalt emits radiation in all directions, a massive spherical head is required to house the shielding necessary to block the unwanted gamma radiation. Only a variable opening for a beam directed at the tumor is left open. The counterweight counterbalances the head and also acts as a beam-stopper to protect other personnel from the beam passing through the tumor.

In 1959, the Theratron was redesigned to reduce material costs. The expensive but compact tungsten shielding in the head was replaced by the more cost effective lead and the counterweight shielding was replaced with cast steel. The cost benefit was offset by the resulting increase in volume of both the head and counterweight, which threatened to make the machine look monstrous. McIntosh's chief responsibility was to create a pleasing and integrated form that was accomplished by designing a shape that flowed from head through a tapered vertical arm to the rectilinear counterweight, leaving ample room for the operator to access the stretcher and patient.

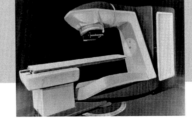

Product Rendering

Theratron 80,   1976

The labour intensive grinding, polishing and multi-coat painting of the rough steel castings of the Theratron 60, was becoming cost prohibitive. A less expensive internal steel structure made of welded steel components was proposed. McIntosh, who had been busy with other AECL projects since 1959, was asked to implement this approach and cover the "ungainly" structure with a pleasing vacuum formed plastic (ABS) cover. A serendipitous result was that all internal components were easier to access and maintain with the removable covers, than with the previous machine, where access was through small holes in the casting.

Atomic Energy Canada

Linear Accelerator
1978
Therasims (Therapy Simulator)
1974

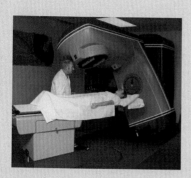

Linear Accelerator,   1978

Linear Accelerators, which use much more powerful high intensity X-rays for treatment rather than cobalt, are replacing cobalt machines. The X-rays are generated electrically and can be turned on and off and focused magnetically, unlike gamma rays. Since they are directional and concentrated on the tumor, they do not need the massive omni-directional shielding, nor do they need a counterweight. Nevertheless, the size of internal components dictates a large and bulky exterior.

In order to integrate the total form and reduce the apparent mass, a linear moulding was introduced where the inside and outside ABS plastic covers meet.

Therasims (Therapy Simulator),   1974

Designed for setting up patient parameters for later treatment on therapy machines, using a light source to simulate gamma or X-rays. The rectilinear form derives from the parts being able to telescope into each other allowing for multidirectional adjustability.

Zeiss

Scanning Electron Microscope
1980

Product Rendering

The electron microscope is a step beyond the most powerful optical microscopes, with magnification up to 100,000 times. The desk was designed to embody all the necessary components, with the microscope itself on the left and the control console on the right, which were ergonomically designed and logically sequenced.

Avmor Ltd.
Dripmatic, 1965

The Dripmatic is a wall mounted deodorizer dispenser for men's' urinals, which operates on a wicking principal. The polypropylene housing is a one piece moulding with an integral hinge that simplifies servicing and eliminates the corrosion caused by the original metal containers. An open grille front permits some of the disinfectant vapors to pass through to nose level, while the sloping top prevents the placing of burning cigarettes on it.

Owl Instruments
Exerpacer, 1975

Invented by Dr. Organ of Owl Instruments, this device made of electronic components activated by a simple nine-volt dry cell battery, indicates pulse rate from 40 to 220/minute with 98% accuracy, and is used primarily after exercising to determine if the workout is having a meaningful effect.

McIntosh and Associates designed an easy-to-assemble, aluminum extrusion that houses the electronic components that fit both the client's budget and production run. Comfort grip handles attached on either end of the housing, will register a pulse when they are simultaneously gripped by both hands.

Danfoss Manufacturing Co.
Tempress Shower Valve, 1967

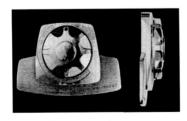

Canada's 1st fully automated shower control maintains a steady mix of hot and cold water to within a ½ degree Celsius, by using a pressure control valve that compensates instantly for any variation in supply pressures. Industrial design involvement for McIntosh and Associates here was focused on the escutcheon and handle control dial, with emphasis on ease of turning and positioning. Universally designed to be operated by children and the elderly, the dial is rotated 180 degrees from the "Off" setting to the "Warmer," setting, with the temperature being indicated by a series of dots rather than a number dial.

Recreational

McIntosh

The manufacturer of the amphibious all terrain vehicle, the ARGO, needed a track to fit over its existing tires for certain ground conditions. McIntosh designed the track section to provide extra traction while avoided the familiar thwacking sound as each section hit the ground by giving the face of each section the same radius as the tire so that the section rolled onto the ground instead if striking it.

Canadian Patent Nº:1215735

An add-on track system for an all-terrain vehicle is disclosed. ATV tires are very soft and flexible, and apt to climb out of a track if the track is not properly located. The invention has a track segment with arms protruding up from the sole-plate of the segment, and the arms converge inwards. The upper ends of the arms locate the segment against the tire side walls, but at a level on the sidewalls where the tire profile is free of the bulging under load which is so marked in ATV tires. The segments have holes through the sole-plate, through which the overhanging parts of the converging arms can be moulded, without movable cores. Retractable cleats occupy these holes, and are forced out when the tire bulges under load.

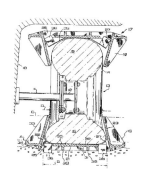
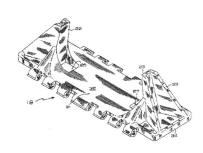
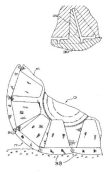

Plate 12

CONCLUSION

If anyone were to ask me if I have missed professional design since retirement I would have to say no. Much as I delighted in the activities at the time, for several reasons I am content to be free from the responsibilities:

1. At age 70 it was probably time to retire.

2. I enjoyed applying my acquired expertise and talents manually; and now such drafting and rendering expertise is supplanted by the computer. I am sure this would have detracted from my enjoyment of the design experience.

3. Today, more and more, if not most products are electronically operated, sometimes including a computer chip, and electronic packaging is not the most stimulating kind of design activity. Cell phones, remote unlocking, starting and other wireless devices are uninspiring, from a design perspective – little black boxes.

4. I am put off by frivolous trends and movements that tend to pervert design activity from the real systematic criteria peculiar to any specific project. I realize that a designer must be aware of current market trends and cannot design in a vacuum; but again, satisfactory performance is key.

5. Most small appliances (e.g. the coffeemaker) and many other products are now manufactured off shore and design cannot be far behind. As any economy matures labour costs escalate. Even Japan and Korea are suffering from such trends. Examination of new products turns up unexpected and even unfamiliar countries of manufacture.

Yes it is pleasant to be free from the responsibilities, to look back on a successful career and to know that one's efforts were appreciated.

Printed in the United States
By Bookmasters